The Language of Birds

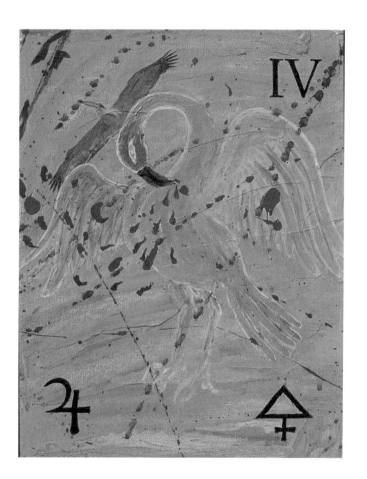

The Language of Birds

of Birds

Some Notes on Chance and Divination

Dale Pendell

FOREWORD BY ANDREW SCHELLING

Three Hands Press

2021

Cover Image: *Blue Cranes*, by Dale Pendell.
Frontispiece Image: *Mitleid* by Dale Pendell.

Foreword © copyright 2020 Andrew Schelling, all rights
reserved.

Book design by Joseph Uccello and Daniel A. Schulke.
Typeset by JFU.

ISBN-13: 978-1-945147-31-9 (softcover)

Printed in the United States of America.

www.threehandspress.com

CONTENTS

Dedication

These reveries were in large measure inspired
by my walks with Norman O. Brown. We
talked often of chance. For Brown, chance was
Dionysus, and a challenge to Divine Order. I saw
chance as a metaphysical refuge, and fought,
squirmed, and argued until the last defensive
bulwarks grew wings and took flight.

This is dedicated to you, Norman O. Brown,
1913–2002.

Foreword

*D*O not be misled by the modest girth of this book. It is a huge volume. A field guide to the future, to fortune, luck, destiny, prediction, chance, randomness, predestination, fate, call them what you like. Pendell's pocket-size volume serves as a guide to all that might come to pass, as well as what might not. Dale Pendell was one of North America's last encyclopedic scholars of the persecuted sciences of prophecy, plant allies, divinatory poetry, and mystical subcultures. Here he stares into the pool of divination, and examines how one might approach the future. It is an ancient art, often placed in the hands of a shadowy figure: Lady Luck, Goddess Fortuna, Madame Roulette—but those grand ladies are really veiled names for the greatest uncertainty. An uncertainty Pendell courted much of his life. This book pins a steady eye on how to roll the bones and gaze upon the dead certain, the hopelessly uncertain, and the utterly impossible.

Behind every divinatory practice lies the premise that the world is fashioned of symbols. That these symbols form a language a person can read. Or misread. It seems apt that Dale moved to the Sierra Nevada foothills in June, 2003, with his wife Laura. They named their homestead Mantis Hill for a noted figure of the local ecosystem, the praying mantis, often found by the barn they transformed into a deep, cool, dark library. They made handwritten labels for their book collections: Prophecy, Plants, Critters, Psychedelics, Hermeticism, Zen. The library became the site of divinatory thought, with *Mantis religiosus*, the praying mantis, standing watch, its spiky green or brown forelegs folded.

The ancients thought this insect watched the future. They gave it the Greek name for prophet.

Which is to say Dale's prophetic tradition, guarded by a biological creature, sits within an ecosystem of live oak, poison oak, coyote, turkey, fox, squirrel, manzanita, and wildfire. I suspect the original job of the mantis was to be diviner of the ecosystem.

Dale's special lifelong study was botanicals. A sizeable portion of his library holds books on plants. Prophecy for him must have settled close to the world of *simples*: medicinal herbs and medicines compounded of herbs, oils, clays, and animal substances. Among plants Dale held a special affection for what he called the *diviners*: peyote, datura, psilocybin mushrooms, the powder blue deceptively gentle-looking flower of the morning glory. *Salvia divinorum*, diviner's sage, held the rank of highest honor in Dale's ecology of mind.

Dale Pendell wrote this book right on the heels of another, *Walking with Nobby*, about his friend and fellow philosopher Norman O. Brown. Brown had once filtered all of philosophy down to a single inquiry: "The basic question is whether chance is blind." You could reword this to ask, "Can the future be seen?"

Brown spoke of his own writings as taking a form "so perishable/that it cannot be hoarded by any elite/or stored in any institution." He did this by using fragmentary voices to seek a harmony of vision. Pendell, with his anarchist cock of the head towards institutions, also employed the use of many voices, but he set those multiple voices to mock, insult, contradict, and knock each other off balance. Not multiple voices to advance a single way of thinking. Multiple voices to reveal both what will come to pass and what will not, what is for sure, and what could never be. Dale set out to wreck the law of non-contradiction. He found a language he wanted in oracles, in the Pythoness, the one whose words carry mixed messages.

This could drive you mad, many voices coming at you, sly, taunting, satirical; one moment agreeing with each other, the next bickering, like the California Indian medicine powers Jaime de Angulo described and which Dale thought about carefully. Friends would now and then caution him, nervous that madness lies in that direction. But Dale knew his Greeks, their thought and their etymologies. He had read E. R. Dodds: "The association of prophecy and madness belongs to the Indo-European stock of ideas." Both come from the same etymological root. Mancy is mania.

The Language of Birds is about reading signs and symbols. It is about recognizing that things are symbols of themselves, but also symbols of other things. Everything points to something not itself. This would be a good first definition of ecology. The skilled outdoorsman, hunter, or scientist watches and listens to the birds, since their calls, or their silence, disclose the presence of other animals, particularly predators. Hawk, weasel, grey fox, mountain lion. Even no voice at all gives powerful counsel.

Notice the list-poem that forms the backbone of this book. An alphabetical string of "-mancies," methods of divination. During the European Middle Ages scholars catalogued oracular practices, listing them in *grimoires*, books of magic. If the word sounds like "grammar," well, that's where it comes from. All magic depends on how you read.

Names for divinatory practices carry two endings. Those thought intuitive or mystical take the ending *-mancy*. Those based on science receive *-scopy*, which means to observe. Some practices take either ending. You can practice hepatomancy or hepatoscopy: predictions based on reading the liver of a slain or sacrificed animal. *Hepatomancy* appears in Dale's list. He passes over it without comment, which may or may not mean he read deeply into it at the time. A few years after this book's publication, Dale received a diagnosis of liver cancer, probably from hepatitis which he had carried for decades. He underwent a liver transplant, emerging from the ordeal a skilled painter as well as writer until his death in early 2018. What did he see those many years spent reading his own liver? Divination is not to be toyed with. It concerns itself with intricacies of life, love, death, and the unmarked paths between. The Goddess Fortuna gives up her secrets reluctantly. This book shows how many voices she uses to do it.

—Andrew Schelling

Preface

Chance favors numerous habits:
flippant, fortuitous,
hap and portent,
uncertain waver,
ultimate author,
risky ally,
fateful nemesis.

Favors, chance favors, fortune favors
the bold, the prepared.

"Alas, m'lord, by chance…"

Perchance,
on a stochastic fulcrum,
Divine Aim:
desultory cadenza,
dense song,
a shuffle dance crane-wrought
in ominous glyphs—
a pachinko telos
cascading from a hand
with 2.718 fingers,
or a ghostly rebellion
against the stacked deck of privilege.

Prayers incline her way,
kneeling suppliants

betting on a knucklebone revelation.
Casual, causal
(it depends on us):
a lucky fall.

Chance is the accidental liberator of heaven,
an apocalyptic alternative cast by lot,
the occult avatar of nihilistic fair play,
immortal threat to eternal order.

Ground of existence.
Hope for newness.
Smile of mantis.
The last excuse and the final request.
Necessity is her twin.

THE LANGUAGE OF BIRDS

T He serpent in the tree, offering knowledge.
    Mercury's snakes, the Hermetic power: the interpretation of signs, that which poisons Single Vision.

The Dragon kild by Cadmus is ye subject of our work, & his teeth are the matter purified. Democritus (a Grœcian Adeptist) said there were certain birds (volatile substances) from whose blood mixt together a certain kind of Serpent (☿) was generated wch being eaten (by digestion) would make a man understand ye voyce of birds (ye nature of volatiles how they may bee fixed).

St John ye Apostle & Homer were Adeptists.

Sacra Bacchi (vel Dionysiaca) instituted by Orpheus were of a Chymicall meaning.

—ISAAC NEWTON

The serpent as the bearer of telluric power. Both Cassandra and Helenos received their prophetic gifts from a serpent that licked their ears, enabling them to understand the language of birds. Siegfried ate the heart of a dragon.

Hebrew prophecy came from snakes:

נְחֻשְׁתָּן

NEHUSHTAN,

the bronze serpent that Moses affixed to a cross.

נְחַשׁ נְחֹשֶׁת

NÂCHÂSH NECHÔSHETH,

serpent of bronze. Both words from

נָחָשׁ

NÂCHASH,

"to hiss, whisper, to divine."

משיח=נחש

=358

NÂCHÂSH=MÂSHÎYACH:

the Serpent is the Messiah.

Guard the Mysteries!
Constantly reveal Them!
—LEW WELCH

THe art of reading signs is one of our most ancient traditions, and a specialty of our Guild. It is not the path to happiness, they say, but what are the choices?

The basic question is whether there is meaning to co-incidence.

The basic question is whether chance is blind. The basic question, the question, is that of divining, glimpsing, seeing forms in chaos.

The matter of augury.

> *The whole outward world with all its being is*
> *a signature of the inward spiritual world.*
> —BOEHME

AEROMANCY
divination by weather or by throwing sand
into the wind

ALECTRYOMANCY
divination by roosters pecking grain

ALEUROMANCY
divination by flour or messages baked in cakes

ALPHITOMANCY
divination by barley

AMBULOMANCY
divination by walking

AMNIOMANCY
divination by the caul of a newborn infant

ANTHRACOMANCY
divination by watching a burning coal

ANTHROPOMANCY
divination from human entrails

ANTHROPOSOMANCY
divination from facial or bodily characteristics

ARITHMOMANCY
divination by means of numbers

ARMOMANCY
divination from the shoulders

ASTRAGALOMANCY
divination by knuckle-bones or dice

ASTROMANCY
divination using the stars, astrology

AUSTROMANCY
divination or soothsaying from words
in the winds

AXINOMANCY
divination by heating or throwing an axe

Divination forms a continuum, but we could say that at one pole there is "possession," and at the other "reading." By possession we mean that a god or some other spirit enters one's body and takes control—voice, gestures, words—all belong to the god. Reading is interpretive—that all the flowing occurrences of this world are a stream of messages. Somewhere in between, half possessed by fire, half swimming in a sea of total significance, there is inspiration.

Thus the seduction. Thus we eat. Thus we drink our mantic syrups. Nanabozho in the forest. Charlie Parker. Eric Dolphy. Tung Shan's Bird Path: extended hands that leave no trace.

A Fork in the path: one way leads to an image of the world as a book, as a riddle, written in code, each occurrence a presage and glyph of the whole. The other way leads to randomness, mere chance, forever beyond our grasp, casting a shadow of nihilism on an accidental universe. Either way, theology is unavoidable. But in the latter case the language is geometry and statistics, while in the former it is luck and power.

Or is that backwards?

A fork in the path: one leads to, well, not clear, but along the way we dismiss accidents without ado.

> *...were I superstitious, I should see an omen in this incident, a hint of fate...Of course, I explain the incident as an accident, without further meaning.*

> —SIGMUND FREUD
> *Psychopathology of Everyday Life*

Thus spake the great seer.

The other way is through the grove of Fortuna, first-born of Jupiter, the goddess of chance who keeps her own counsel. The other way. The other way leads to the Pythoness, the Oracle.

A Ugury, divination from the flight of birds. Latin *augere*, "to enlarge, increase."

Auis, "bird," is a different word, but Plato puns them. Socrates explains to Phaedrus that it is augury that, through *oiōnos* (a bird of omen), supplies *oiesis* (human thought) with information and "mind" (*nous*).

The *oionoistic* art: the reading of signs, that salient peculiar-
ity of our species.

> *Bird-reckoning. Bird-brain.*
> *The bird as the authority:*
> *finding august signs, images, white flecks in liver.*
> *Or coincidence of words:*
> *nicknames, or the waxing moon.*

Auspicious increase. The bird-spectre speaks.

> *It furthers one to cross the great water,*

to eke out meaning.

> *Put two ounces of lead into a steel ladle or a cast-iron frying*
> *pan and melt it on the stove. When it is molten pour it into a*
> *can of water. Your fortune can be read from the shape of the*
> *solidified lead. Do this on New Years Day.*

BELOMANCY
divination by marked arrows

BIBLIOMANCY
divination by random Bible passages (pagans
preferred Homer or Virgil)

BLETONOMANCY
divination by ripples or patterns in
moving water

BOTANOMANCY
divination by plants

CAPNOMANCY
divination by smoke, or bursting poppy heads

CARTOMANCY
divination by cards

CATOPTROMANCY
divination by a polished shield or mirror

CAUSIMOMANCY
divination from the ashes of burned leaves
or paper

CEPHALOMANCY
divination by a boiled donkey or human skull

CERAUNOSCOPY
divination by lightning and thunder

CEROMANCY
divination by molten wax poured into water

A Ugury, divination: Socrates distinguishes between the two, claiming that divination is the higher art because it comes from divine madness, and is thus a gift of the gods, whereas augury is the pursuit of rational men.

Tall Hector, eyes averted under his flashing helmet, shook the two lots hard.

But the difference is not so clearly cut. Visions are also signs, as are voices. How much of what we ever see is not projection, or hallucination? Is vision a mental or a physical phenomenon? If you see it in darkness, or with your eyes closed, is it still seeing?

One of the Principal Ones spoke to me and said: "María Sabina, this is the Book of Wisdom. It is the Book of Language. Everything that is written in it is for you. The Book is yours. Take it so that you can work."

—ALVARO ESTRADA
María Sabina, Her Life and Chants

D Ivination is one of the most common practices associated with the use of entheogenic plants. Peyote and datura, mushrooms and morning glory seeds, all were used for divination in Mesoamerica. The entheogens could as correctly be called the Diviners.

[Peyote] causes those devouring it to be able to foresee and to predict things; such, for instance, as whether on the following day the enemy will make an attack upon them; or whether the weather will continue favorable; or to discern who has stolen from them some utensil or anything else; and other things of like nature which the Chichimeca really believe they have found out.

—FRANCISCO HERNANDEZ
De Historia Plantarum (in La Barre)

Mushrooms. Tobacco. Coca. *Ololiuhqui. Salvia divinorum.* And the empathogens.

CHAOMANCY
divination from the appearance of the air

CHARTOMANCY
divination from written pieces of paper

CHIROMANCY
divination by the nails, lines, and fingers of the hand

CHRESMOMANCY
*divination from magic sounds or
foreign words*

CLAIGUSCIENCE
*divination from the taste or smell of a food
that is not present*

CLEDNOMANCY
divination from hearing a chance word

CLEIDOMANCY
divination by a suspended key

CLEROMANCY
divination by the casting of lots

COSCINOMANCY
divination by a sieve suspended on shears

CRITHOMANCY
*divination by grains sprinkled on burnt
sacrifices*

CROMNIOMANCY
divination by onions

CRYSTALLOMANCY
*divination by crystal ball or the casting of
gemstones*

CUBOMANCY
divination by throwing dice

CYCLOMANCY
divination by the wheel of fortune

A Uspice. Auxin: the growth hormone in plants. Sanskrit *ukṣati,* "he grows." Plants as authors, the way of growth and spontaneity, across Central Asia, strengthening, maybe getting fat, a big waistline, until we grow old among the Tocharians: *okṣu.*

L Atin: *divinatio,* related to *divinare,* "to predict," and to *divinus,* "divine," "pertaining to the gods."

Greek: *manteia,* "divination." A prophet or prophetess is *mantis,* related to *mainomai,* "to be mad," and *mania,* "madness," all from the Proto-Indo-European root **men.*

> *If the Greeks were right in connecting* mantic *with* mainomai—*and most philologists think they were*—*the association of prophecy and madness belongs to the Indo-European stock of ideas.*
>
> —E. R. DODDS
> *The Greeks and the Irrational*

*Men, "mind," is also the root for "meaning." Thus there is meaning in madness.

According to Homer, the mantis was always welcome at a prince's table, along with carpenters, doctors, heralds, and poets.

Rationality. Ratio. Analysis. Pascal's adding machine: stacks of Boolean gates. Computers can beat grandmasters: it's clear that logical deduction is not our particular forte.

Madness may be.

The Greeks had two words for chance, *tyche,* and *automatia.* Aristotle used *tyche* to refer to coincidental occurrences in the human world, and *automatia* to refer to chance events in the natural world, the difference being that human beings possess free will while natural objects do not. For Democritus, *automatia* refered to events that had no external cause, while *tyche* meant that it was an event for which we were simply ignorant of the cause. "Hidden variables." Democritus didn't believe in randomness.

Randomness is entropy, Claude Shannon's measure of information. Patternless chaos is maximal meaning, incompressible and incomprehensible.

$$-(\textbf{entropy}) = k \log (1/D)$$

We don't have the encryption key.

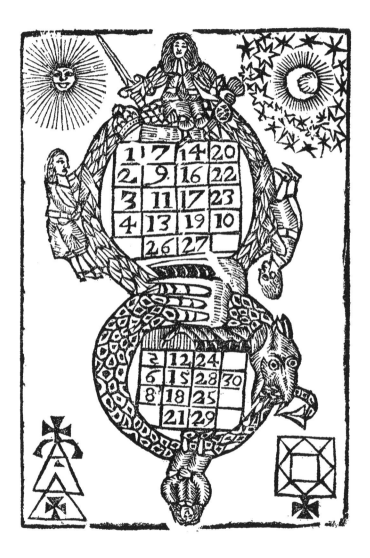

Fortuna as the robot genius. Self-moving. Self-willed, and (perhaps an error here) self-contained, uninfluenced by the environment. Insulated from stray sparks.

Who but a robot can be spontaneous?

Automatons are glamorous: enchanting embodiments of grammar. Incarnation of the Logos.

The primordial robot was the Big Bang—Democritus's first cause and Timolean's goddess of chance:

Αυτοματια, Automatia.

Or maybe the Creator is a marionette. Or perhaps there is no such thing as chance.

There is always that chance...

A polysemous world: the path suffused with numinosity, a quavering feeling that everything is portentous, happening as if by prediction. A hazy presage, like a mirage. Pin it down and it disappears. A crow. A wild duck.

Where did it go?

-mancy. From μαντις, diviner. Sanskrit *múnis*, a seer, or sage. Perhaps related to *mátis,* "mind."

Thus Shākyamuni, the Buddha, is the "seer of the Shākyas."

Tibetan diviners were called *mopas*. In addition to oracular prophecy through trance and possession, Tibetans practiced augury with birds, dice, arrows, mirrors, and the rosary *(mala)*.

> *All these [divinatory] operations, in the world of psychical phenomena as in the world of physical phenomena, may be carried out either in awareness of their relative and ultimately illusory nature and with regard to their moral consequences, and therefore in compatibility with the teachings of Buddhism, or without that awareness and regard.*
>
> —LAMA CHIME RADHA RINPOCHE
> (in *Oracles and Divination*)

Divination is used to find which of the five spiritual families of practice best suits an initiate.

DACTYLIOMANCY
*divination by suspended finger ring or
pendulum*

DAPHNOMANCY
*divination by the crackle of roasting
laurel leaves*

DEMONOMANCY
divination with the help of demons and spirits

DENDROMANCY
divination by oak and mistletoe

ELAEOMANCY
divination by the surface of water

ENOPTOMANCY
divination with a mirror

FELIDOMANCY
divination from the behavior of wild cats

GASTROMANCY
divination by food, or sounds from the stomach

GELOMANCY
divination from laughter

GEOMANCY
divination by cracks or lines in the earth, or dots on paper

GRAPTOMANCY
divination from handwriting

GYROMANCY
divination by spinning in a circle until dizzy

O Rder is syntax, συνταξις: the formations of soldiers, order of battle.

When there is conflict, the masses are sure to rise up.
Hence there follows the hexagram of THE ARMY.

The grammatical army: Chomsky's context-free grammar. The Turing machine as the neurolinguistic engine of war, more like Lamarck than Locke, etched into codons, DNA. Or BNF: Backus-Naur Form.

         \<Exp\> ::= \<ident\>
                | \<constant\>
                | l \<ident\> . \<Exp\>
                | \<Exp\> \<Exp\>
                | (\<Exp\>)

Chomsky thinks that we're born with it.

Arms are used when there is no other choice; this
is like using medicine to cure illness. Therefore
it is called "poisoning the country."

—CHIH-HSU OU-I:
I Ching Commentary

That the Von Neumann wave function collapses implies order: observables emerge, an implicate order, as in Bohm, or the *Avatamsaka*—a non-syntactical order, holographic, each part containing the whole.

Non-sequential organization: John Cage (Gaussian), Norman O. Brown (corporeal), Gary Snyder (dendritic). Or James Joyce. Open the book at random. Ordered by mandala, rather than Cartesian axes: no origin, like Anaximander's cylinder. *Apeiron.*

Alan Turing's oracles were deterministic, and therefore not mad, and, as Roger Penrose shows, following Gödel's proof, incapable of understanding. They can't solve the halting problem. Penrose suggests that a non-computational brain might need a quantum time loop, so that the results of future computations are available in the present.

> *Thus all beings have thought by the will of chance.*
> —EMPEDOCLES

The monkeys are still at the typewriters, working on *Hamlet.*

HALOMANCY
divination with salt

HEPATOSCOPY
divination by the liver of a sacrificed animal

HIEROMANCY
divination by interpreting sacrifices

HIPPOMANCY
divination by the behavior of horses

HYDROMANCY
divination by water or tides

ICHTHYOMANCY
divination from the movements or entrails of fish

IDOLMANCY
divination from movie or rock stars

LAMPADOMANCY
divination by the flickering of torches

LECANOMANCY
divination by looking at oil or jewels in water

LIBANOMANCY
divination by staring at the smoke of burning incense

LITHOMANCY
scrying with gemstones and natural crystals

LOGARITHMANCY
divination by logarithms

LYCHNOMANCY
divination by flame of an oil lamp or candle

C Hance is creative, a subtle propensity for change. Swerve, the *clinamen* of Epicurus and Lucretius. An inclination. Grain in wood. Desert varnish on sandstone. The affinity of things.

> *The fact that life disturbs the order of the world means literally that at first, life is turbulence. What you see from the top of the cliff, in its sweetness, is the first-born being arising out of the waters, Aphrodite, who has just been born in the swirl of liquid spirals, Nature being born in smiling voluptuousness.*

> —MICHEL SERRES
> *"Lucretius: Science and Religion"*

Voluptas or voluntas? The critical letter is obscured in the only extant copy *of De Rerum Natura.*

Serres states it beautifully:

> *Starting from Venus, the natural, projected to society, gives materialism; starting from society, that, projected to science, gives hierarchy and determinism.*

It's the *V* of Venus, the angle, that creates turbulence in a laminar flow: chaos overwhelms causality.

    The Epicureans called augury superstition, but the blindness/vision of the *clinamen* is still being debated in physics. How far does quantum entanglement go, and does it mean anything at all?

> *Suam habet fortuna rationem* (Chance has its reasons).
>   —PETRONIUS

She loves me, she loves me not.

Dice seem unnecessary, and slightly vulgar: reason is chancy enough. Closer to Planck's h. The brain must be indeterminate, if the wave function is, so intellect is Ψ, an aleatory endeavor.

Is it a sign, or had you best trust to reason?

Everything *we* do is chance. Only automatons can create order. Unless, as Ouspensky pointed out, we are really machines.

Automatic means not moved by external causes: self-willing and self-creating. Self-actualizing.

Edward Grieg prohibited his family from humming or singing any but his own melodies, so that his compositions would not be "influenced."

Chance may or may not be deterministic. Chaotic dynamics are deterministic but unpredictable. Computers generate pseudo-random numbers indistinguishable from truly random numbers. What truly random numbers? Perhaps we need a Turing test: if a series of digits cannot be distinguished from random digits, they *are* random digits.

Philosophy is so deadly. How can you read it? How can you listen to it?

Randomness becomes a tautology, circular: "true" random digits come from radioactive decay, random by the definition of Ψ^2.

But physics is divination, arithmomancy. There is no frictionless surface, and there is always a third body.

Chance vs. order.
(Chance as chaos, anarchy loosed on the library.)

Chance vs. teleology.
(The atheistic Liberator: the Enlightenment.)

Chance vs. determinism.
(The problem of Free Will in a pinball machine.)

Reading signs is not the path of happiness. We kill the messenger. It was a capital offense to prophesy the death of the Emperor.

Teiresias was blinded,
Phineus also;
Amphilochos slain.

Prometheus chained to a lonely rock.

MACHAROMANCY
divination by knives or swords

MACULOMANCY
*divination from the shape and placement of
birthmarks*

MARGARITOMANCY
divination by heating and roasting pearls

METEOROMANCY
divination by storms and comets

METOPOMANCY
divination by examining the face and forehead

MOLYBDOMANCY
divination by dropping molten lead into water

MYOMANCY
divination by squeaks of mice

NECROMANCY
divination by ghosts or spirits of the dead

NEPHELOMANCY
divination by appearance of clouds

NIGROMANCY
*divination by walking around the graves of
the dead*

L Ady Luck, the gambler's god. If there is any truth to it at all, there must be gamblers who win.

The signs are favorable

Fortuna loved lots, and playing dice.

Or is Fortune a Wheel? In early tarots, such as the fifteenth century Visconti deck, Fortuna is blindfolded, as is Cupid on card VI. Love and fortune, partners in divine madness.

From ancient times, Fortuna was associated with Anagkê, Necessity.

F Ortuitous, *fors, fort.*
     Fortitude. Power.
     But power is always chancy. The helping spirits may not come, they may be busy—they may be playing their own bone game, and they may not want to interrupt it just to help out some mortal shaman.
     The work is finding lost objects, or a lost shadow, discerning the cause of an illness, or a streak of bad luck.

Without power you cannot do anything out of the ordinary.
With power you can do anything. This power is the same thing

as luck. The primitive conception of luck is not at all the same as ours. For us luck is fortuitousness. For them, it is the highest expression of the energy back of life.

—JAIME DE ANGULO

And luck is connected with wildness. *Mana.* Coyote has *mana*, but screws up anyway, thus ensuring that nothing goes according to plan. Coyote is kind of like chance.

"Surely, Professor Bohr," asked a visitor to the physicist's country cottage, "you do not really believe that that horseshoe over your door brings good luck."

"No," answered Bohr, "I certainly do not believe in superstition. But I have heard that horseshoes bring good luck even to those who do not believe in them."

C Hance as the *deus ex machina.* The Final Metaphor. It explains everything, and nothing, thus called "mere."

The Goddess reduced to insignificance at the very moment of her supreme authority.

CHANCE VS. MEANING.
(It's just chance, darling.)
CHANCE VS. WHOLENESS.
(Bohr vs. Bohm.)
CHANCE VS. LUCK.
(Rationality vs. Superstition.)

☿

C *Hance perturbations. Normal distributions.*
There is no such thing as the "zone," or the "streak," according to Gilovich, Vallone, and Tverksy ("The Hot Hand in Basketball: On the Misperception of Random Sequences"). Looking at a year's worth of statistics on the Philadelphia 76ers, everything fits into a normal distribution. (In baseball, Joe DiMaggio is still a problem.)

The "clustering illusion," the human tendency to see patterns where none exist. Related to *pareidolia*, the illusion of control, the gambler's fallacy, the Texas sharpshooter fallacy, and perhaps to eighty other named and described cognitive biases, all irrational. If we add Freud to that does reason have a chance?

But maybe they've just proved that luck is modest, that she disguises herself with bell curves, that she loves static and noise. The shuffling and the shaking, the randomizing, was to eliminate human bias, that the Word of God be known.

Or maybe luck is power, as in "being in tune with." Democritus's "eidola," or maybe Lao Tzu's "tao."

No victor believes in chance.
—NIETZSCHE

THe "clustering illusion." Seeing patterns where there are none, or, since Frank Ramsey has proved that compete disorder is an impossibility, giving meaning to patterns that have none. As rational men of the Middle Ages sought to debunk the superstition of fortune with the Truth that all things are governed by God, rational men of today wish to debunk the irrational belief in luck with the Truth that all things are governed by chance. We're back to Empedocles.

And the scientists are as smug as the churchmen.

At least the churchmen knew they were
dealing with theology.

P *Areidolia*: a vague and random stimulus being mistakenly perceived as recognizable, such as seeing animals or faces in clouds, or a man in the moon.
Or constellations. Or a face in a face. Or a pipe.

This is a random sequence of letters and spaces.

And certainly, most certainly, most rationally,
no hallucination.

Soldiers tend to embrace Fate, or Destiny: "if it's not your time, you won't be killed."

Fate is static, beyond even the power of the gods. Divination implies indeterminacy. If the future were fixed, what is the point of the consultation?

Chance vs. Choice

If you don't choose, you'll go on ending up with just whatever comes by.

(Yep. Sigh. But maybe the World has more intelligence than I do.)

As the Epicureans scoffed at divination, even when it proved true, the Stoics defended divination, even when it proved false. The Stoics were determinists, and believed in Fate. Fortune was relegated to subjective moments of incomplete understanding.

WE defy augury.

Alexander went to Delphi before beginning his conquests of the East. The Oracle was closed, there being only a few days each month when the Pythoness did consultations. Alexander pressed his case several times, but the priestess was firm. Al-

exander grabbed her by the hair and began dragging her down the stairs to the Adytum. The Pythoness agreed to perform a consultation and added, "Alexander, you are invincible." Alexander immediately let go of her, thanked her for the consultation, and departed.

During the first Punic War, Publius Clodius was leading an expedition against Carthage. As was customary, before commencing the attack, they sought an alectryomantic oracle from the chickens they kept on the ship in cages for that purpose. In this case, the chickens would not eat the grain that was offered—a bad omen. Clodius ordered them thrown into the sea, saying "if they won't eat, let them drink." His expedition met with disaster. He was convicted of treason and executed—not for failing militarily but for blasphemy.

In Tenochtitlan, the wealthy could pay to have an unfavorable birth date changed.

OCULOMANCY:
divination by observing the eye

OINOMANCY:
divination by gazing into a glass of wine

OLOLYGMANCY:
divination by the howlings of dogs or wolves

OMPHALOMANCY
*divination by counting knots on the
umbilical cord*

ONEIROMANCY
divination by the interpretation of dreams

ONIMANCY
*divination using olive oil to let objects slip
through the fingers*

ONOMATOMANCY
divination by the letters in names

ONYCHOMANCY
divination by polished fingernails

OOMANCY
*divination from drops of fresh egg whites in
water*

OPHIOMANCY
*divination by the coiling and movement of
serpents*

ORNITHOMANCY
divination by the flight or songs of birds

OSTEOMANCY
divination from bones

C Hinese teh, 德 or tê, "virtue," also means "power," in the sense of accumulated luck. Thunder and wind: the image of DURATION.

> *If for example I consult the tortoise and get a favourable re-*
> *sponse, that is my tê. It is my potential good luck. But it re-*
> *mains like an uncashed cheque unless I take the right steps*
> *to convert it into a fu, a material blessing. Like an uncashed*
> *cheque, a tê is a dangerous thing to leave about. It may fall*
> *into other hands, be put into someone else's account.*

—ARTHUR WALEY
The Book of Changes

Written or scratched onto a tortoise shell, or nailed to a tree. When an interpretation is accepted, it becomes a power loose in the world. But likewise an omen may be deflected, may be reinterpreted. Quick wit and a ready response. Alexander in Persia before the sweating statue of Orpheus. The power of Thoth. Or thought.

The Athenians locked their state prophecies away in a temple, lest their enemies see them.

According to Aeschylus, men learned the arts of divination from the one who stole fire from the gods, Prometheus. Earth and fire. Apollo. Sun and earth. Sun and moon. Lunacy.

Augury is the realm between.

Apollo stole the secrets of prophecy from Pan. At Delphi, Apollo and Heracles fought for the tripod. The tripod belonged to Gê, who gave her oracles in darkness, from dreams, onieromancy, or from lots, cleromancy.

> *Serpents were much in evidence at the oracle of Delphi.*
> —M. P. HALL

Most of the questions asked at Delphi were personal, matters of business or health, but the cities consulted her also on every important state decision.

> *All Attica will be taken,*
> *but Zeus grants Athena a wooden wall*
> *Which alone will be untaken.*

Crœsus also consulted the Oracle, sending a caravan of gold overland from Sardis. Philosophers transcribed the Pythoness's utterances, the thespiod put them into verse.

PEGOMANCY
divination by bubbles in springs or fountains

PESSOMANCY
divination by pebbles

PHILEMATOMANCY
divination by kissing

PHYLLOMANCY
divination by the patterns and colors of leaves

PHYLLORHODOMANCY
divination by clapping rose petals between
the hands

PHYSIOGNOMY
*divination by shape, marks, and proportions
of the body*

PLASTROMANCY
divination by tortoise shells

PODOMANCY
divination by the soles of the feet

PSEPHOMANCY
*divination by rolling small stones, or selecting
them at random*

PSEUDOMANCY
fraudulent fortune-telling

PSYCHOMANCY
*divination from the state of the soul, alive
or dead*

PYROMANCY
divination by fire or flames

RETROMANCY
divination by looking over one's shoulder

RHABDOMANCY
divination by branches or rods, dowsing

RHAPSODOMANCY
divination by a book of poetry

THe Pythoness burned laurel leaves—
sweet smelling vapour from the Adyton.

Henbane (*apollinaris*), black and white hellebore, and datura were used at the oracle of Trophonios at Lebadea, the consultants also partaking.

At the oldest oracle, at Dodona, an ancient oak gave voice to Zeus when the wind blew through the branches.

unwashen feet, sleeping on the ground

The first auguries may have come from the behavior of animals.

Serpent, lizard, bat.
Stuttering. Sneezes.

Bird, *oino. Ornis*: portent. Zeus: eagle, vulture. Apollo: raven. Hera: crow.

Cicero thought fish too dumb to speak for the gods.

True dreams, false dreams,
Gate of Ivory, Gate of Horn.

Our most ancient possession.

SCAPULIMANCY
*divination from cracks in a charred
shoulder blade*

SCATOMANCY
divination by studying feces

SCIOMANCY
*divination from shadows or the shades of
the dead*

SCYPHOMANCY
divination by cups or vases

SELENOMANCY
*divination from the phases or appearance of
the moon*

SIDEROMANCY
divination by the burning of straws

SPASMATOMANCY
divination by twitchings of a body

SPATILOMANCY
divination by animal droppings

SPHONDYLOMANCY
divination from beetles or other insects

SPODOMANCY
divination by ashes

STICHOMANCY
divination from random passages in books

STIGONOMANCY
divination by writing on tree bark

STOLISOMANCY
divination by the act of dressing

SUGGRAPHAMANCY
divination by studying history

STERNOMANCY
divination by the breast-bones

SYCOMANCY
divination by drying fig leaves

Without augury there is the general, a ghost story of abstractions: logical order instead of inspired madness. General Cadmus impaled the serpent on a tree and killed it. The teeth grew into soldiers. The Vatican armed.

> *As a general rule among the ancients power was in the hands of augurs.*
>
> —CICERO
> *De Div.* I, 40

> *The original identity of power and wisdom*
> *sapientia, sapience, savor*
> *sapience and divination they considered royal*
> *but if the salt has lost its savour*
> *pellitur e medio sapientia, vi geritur res.*
>
> —ENNIUS

> *in the place of sapience, violence.*
>
> —NORMAN O. BROWN
> *Inauguration*

Cicero concluded that divination and augury were superstitions that perturb the tranquility of mind. But as an augur himself, and a good aristocrat, he allowed that augury was a good way to control what Hamilton called the "excesses of democracy."

A clenched fist raised to the sky.

Augury degenerates into priestcraft, priestcraft into force.

Auspice → Authority.

Unlimited Reason (except it's not) vs. letting the god, or the environment, or the moment, have a say in what happens.

(Except it's not. Except it's not.)

Ivination is chance and chance is playfulness, *lila*, enthusiasm, feeling God within, *entheogen*. Clear portents from a dropped word, from timing. Myriad faces in every tree and bush.

> *Seeing faces in clouds, trees, and such probably has to do with overstimulation of a particular area of the visual cortex. Seeing faces is an especially important skill of vertebrates. If the cortex is damaged, a condition known as prosopagnosia can occur, in which the person can no longer recognize faces as being faces. The reverse condition might be called hyperprosopognosia.*

Excess of meaning, the "too-muchness." A sentence with five levels of meaning, and everybody talking at once. A polymorphous bead game in overdrive: all the meanings anticipated before the sentence is finished, questions and answers, call and response, uttered simultaneously. A diapason of lucid babble: Pentecost.

The early Christians practiced glossolalia. A few still do. Though the Apostles had cast lots, the Church forbade all divination in the third century because of its association with paganism.

> *They drew lots, and the lot fell on Matthias.*

Fortuna was the last survivor. She must have been a bird goddess. She was Etruscan, or earlier, and wore a belt of skulls. The Christians renamed her Providence. Insurance companies cover accidents, but not "acts of God."

Emperors closed oracles the way dictators shut down newspapers.

Apollonius was declared the Antichrist.
Agrippa filled in the cavern of the Oracle at Baia.

L Uck is the context, the environment: set and setting.

Events follow definite trends, each according to its nature.
Things are distinguished from one another in definite classes.
In this way good fortune and misfortune come about. In the
heavens phenomena take form; on earth shapes take form. In
this way change and transformation become manifest.

—I CHING

TASSEOGRAPHY
divination by tea leaves

TEPHRAMANCY
divination by the ashes on an altar

THEOMANCY
divination from the responses of oracles

THERIOMANCY
divination by watching wild animals

TIROMANCY
divination by milk curds, or the holes on cheese

TOPOMANCY
divination by the contours of the land

TROCHOMANCY
divination by wheel tracks

THUMOMANCY
divination by intense introspection of one's own soul

TRANSATUAUMANCY
divination from chance remarks overheard in a crowd

URIMANCY
divination by casting the Urim and Thummin

URINOMANCY
divination using urine for scrying

XENOMANCY
divination by studying the first stranger to appear

XYLOMANCY
divination by wood or fallen branches

ZYGOMANCY
divination with weights

ZOOMANCY
divination by the behavior of animals

David Abram connects the loss of animism with the adoption of the Phoenician/Hebrew alphabet by the Greeks. The letters were no longer associated with animals and objects. When vowels were given signs, writing was separated from its former reliance on oral transmission, and soul (psyche) became associated with the literate intellect. The idea of a pure and detachable soul, transmitted from the Pythagoreans through Orphism and, ultimately, to the Puritans, was rejected by Blake.

> *The Goddess Fortune is the devil's servant,*
> *ready to kiss anyone's arse.*

It was Cadmos brought the alphabet to the Hellenes.

Legends are concocted not without reason.
—THEOPHRASTUS

Reification is idolatry.

ABstracted words are idols. Where the letter stands the spirit has departed. The spirit is the wind: that which surrounds, which nurtures and sustains. The word alone is defleshed, a disembodied intellect. Meaning is not in words. Meaning lies between the lines, between the sheets, embedded in sentence. Or sentience.

Without a body, the alphabet hovers like a hungry ghost above a stagnant well, wailing in an eternal twilight.

Ego is like the centralized state. In the committees, not one member is sober.

Mammalian language. Rooted words. Plants with hair. Without sap no sapience.

As the universe becomes alive, the ruler dissolves and vines grow over the throne. Each member speaks, or sings, a choir, the diapason building. "Arise, ye more than dead!" Undulations move across a field of grasses, rustling, whispering secrets. Somewhere Spinoza is scratching at the window, and the neighborhood enters freely.

THe Hebrews distinguished between true and false prophets: the false prophets not remembering anything when they emerged from the trance, while the true prophets remained conscious. Robert Graves likens the latter to the poetic trance: the words are gifts, from the Other, but the poet maintains full consciousness. The divination is immediate and intuitive, like the Tibetan *tra*, or the second sight of the Scottish highlanders.

But there is another level of the poetry oracle--the internal logic and structure of the poem—its rhythm, assonance, and rhyme—that can demand a word or phrase otherwise quite out of place, semantically. The poem itself discovers, or uncovers, new information, that the poet herself does not know. A poem in resonance is like a formula in physics, an equation of power.

THe great god Pan is dead.

And it can't be blamed on the Christians. There were charges of corruption and political conniving at Delphi. Perhaps the mephitic vapours had ceased wafting through the cavern, and the Adyton was less sweet. Times were changing. It was harder to believe in gods that walked around and drank wine—easier to believe in things like ideas, and numbers.

"It's only…" "It's just…" "But that's irrational." But that's cognitive pathology." "But that's fuzzy thinking." "It's nothing but…"

God, Chance…
     presage for a year of drouth.

Tell the emperor that my finely wrought house has fallen to the ground. No longer has Phoebus his shelter; nor his prophetic laurel, nor his babbling spring: the speaking waters have dried up.

—LAST UTTERANCE OF THE DELPHIC ORACLE

AILUROMANCY
divination by the actions of a familiar cat

ARACHNOMANCY
divination using spiders

EPOMBRIAMANCY
divination from the sound of rain

GLAUXIMANCY
divination using owl castings

HAEMOCAPNOMANCY
divination by the smoke of burning blood-soaked paper tissues

MEDIAMANCY
divination by scanning police radio or random TV shows

OULEIMANCY
divination by the appearance of scars

SELENOSCIAMANCY
divination by the shadows of moonlight through trees

TYMPANIMANCY
divination from the rhythms of drums

TO understand the language of birds, one needs not ears, not cochlea and tympanum, but a cellular hearing, where the organs of perception have expanded to include skin, hair follicles, heart beat, and whatever it is that is all of it together.

The prophetic gift is like a writing tablet without writing, both irrational and indeterminate in itself, but capable of images, impressions, and presentiments, and it paradoxically grasps the future when the future seems as remote as possible from the present. This remoteness is brought about by a condition, a disposition, of the body that is affected by a change known as "inspiration."

—PLUTARCH
On the Cessation of Oracles

There is only poetry.

The second edition of *The Language of Birds* was released by Three Hands Press at the Feast of the Holy Fool, April 2021. It consists of an edition of four thousand ninety-six softcover copies, a limited edition of five hundred hardcover editions in black cloth with color dust jacket, and sixteen hand-bound copies in blue leather with gilt embossing and slipcase. Gilt Lunar Divination Seal on hardcover and special editions by Johnny Decker Miller.

Scribæ Qua Mysterium Famulatur